BEYOND ChatGPT:

Prompt Engineering Journal for Adults

A Guided Blueprint for Mastering Expert Level AI Prompting

MY PROJECT PROMPTS

NAME: ______________________________

DATED: ______________________

THROUGH ____________________

>____<

JOURNAL #______ OF ________

MEMOS:

>____<

>____<

>____<

>____<

HOW TO USE THIS JOURNAL

Welcome to your AI Prompt Engineering Journal for Adults.

This book is your personal studio for building, improving, and mastering the skill of writing effective prompts. Each two-page spread guides you through a repeatable structure you can use for any task - creative, analytical, personal, or professional. Be specific about requirements.

On the left page; clarify what you want, who the AI should pretend to be, the tone to use, constraints to follow, and what format you want the output in. This page helps you think before you prompt and be consistent.

On the right page; write the actual prompt, along with examples, structure, and refinement notes. This page represents the "working prompt" - the one you'll copy and paste into an AI tool. Date and number the versions as you track what works and doesn't work.

Use this journal to:

- Practice clear thinking
- Develop flexible prompt structures
- Document what works and what doesn't
- Refine your skills through experimentation
- Create clear examples and reusable templates
- Track your growth over time
- Group similar prompts together for future "Prompt Stacking"

There are no rules here - only tools. Customize the pages, rewrite them, and use the process that works best for your style. "Talk" to AI as if speaking to an assistant and repeat amended prompts until you get what you are looking for. This journal becomes more valuable with every prompt you write.

1. **Sample Entry #1 - How to Plan a Weekend Project**
2. **Sample Entry #2 - Creative Brainstorming Prompt Example**
3. **Sample Entry #3 - Technical Simplification Prompt Example**
4. **Sample Entry #4 – Business Communication Prompt Example**

SAMPLE ENTRY 1 — Weekend Project Planning

LEFT PAGE

PROJECT / CONTEXT:

How to create a clean simple prompt for planning a weekend project

ROLE / PERSONA:

Creative Organizer and AI Planning Assistant

☑ Expert consultant
☑ Creative writer
□ Data analyst
□ Teacher / educator
□ Other: __________

TONE CHECKLIST:

☑ Professional
☑ Casual / friendly
□ Technical
□ Educational
□ Creative

OUTPUT FORMAT:

☑ Bullet points
□ Numbered list
□ Paragraphs
□ Table / chart
□ Step by step

AUDIENCE:

A busy adult who wants a simple realistic plan

KEY CONSTRAINTS:

No emphasis symbols
Do not repeat instructions
Provide a concise timeline

MEMO:

Use this example to learn how to organize your thinking before writing a prompt. Keep notes short and focused.

SAMPLE ENTRY 1 — Weekend Project Planning

IGHT PAGE

ROMPT TITLE: Weekend Project Planning Prompt
ATE: 02 15 2025
ERSION: 1.0

ASK OR GOAL:

reate a simple Saturday morning plan that lists the three most important steps with ealistic timelines

ONTEXT AND SPECIFICS:
ne user can only work on the project Saturday morning and has limited time. The plan hould feel achievable and not overwhelming.

XAMPLES TO INCLUDE:
ood example:
st the top three tasks with time estimates and keep everything friendly and simple

hat to avoid:
void adding unrelated tasks or providing long unnecessary explanations

RUCTURE OR STEPS:

1. Identify main goal
2. List top three tasks
3. Add time estimates
4. Add one optional task
5. Suggest simple next steps
6. Keep instructions short
7. Maintain friendly tone
8. Follow constraints

CTUAL PROMPT:
ct as a Creative Organizer and AI Planning Assistant. Create a simple Saturday morning lan with three main tasks and estimated time required for each. Include one optional isk if time remains. Keep tone friendly and professional. Do not repeat instructions. eep the timeline concise and easy to follow.

ESULTS OR NOTES:
/hat worked: Shorter prompt produced cleaner output

/hat didn't: Too many optional tasks created clutter

efinements for next time: Add clearer time windows and define priority levels

✨ SAMPLE ENTRY 2 — Creative Brainstorming

LEFT PAGE

PROJECT / CONTEXT:

Brainstorm a list of creative product ideas inspired by nature

ROLE / PERSONA:

Creative Innovation Partner

☑ Expert consultant
☑ Creative writer
□ Data analyst
□ Teacher / educator
□ Other: ___________

TONE CHECKLIST:

□ Professional
☑ Casual/Friendly – (Warm)
□ Technical
□ Educational
☑ Creative

OUTPUT FORMAT:

☑ Short lists
□ Numbered list
☑ Paragraphs
□ Table / chart
□ Step by step

AUDIENCE:

Artists designers or makers looking for inspiration

KEY CONSTRAINTS:

Use beginner friendly language
Avoid overly technical descriptions

MEMO:

Use this example for idea generation and creative exploration. Focus on textures shapes or natural systems.

SAMPLE ENTRY 2 — Creative Brainstorming

RIGHT PAGE

PROMPT TITLE: Nature Inspired Product Ideas

DATE: 02 15 2025

VERSION: 1.0

TASK OR GOAL:

Generate ten unique product ideas inspired by elements of nature

CONTEXT AND SPECIFICS:

Ideas should connect natural textures, shapes, colors or systems to the design or function of the product

EXAMPLES TO INCLUDE:

Good example:

A lamp with a curled leaf silhouette made from translucent green resin

What to avoid:

Avoid generic products that do not clearly reflect a nature based influence

STRUCTURE OR STEPS:

1. Choose a nature theme
2. Identify a pattern or texture
3. Describe function of product
4. Connect nature element to design
5. Keep descriptions vivid
6. Use accessible language
7. Maintain creative tone
8. Stay within ten items

ACTUAL PROMPT:

Act as a Creative Innovation Partner. Generate ten unique product ideas inspired by nature using textures, shapes or natural systems. Describe each concept clearly and explain the nature element behind it. Keep language beginner friendly and creative.

RESULTS OR NOTES:

What worked: Using textures created more vivid ideas

What didn't: Some ideas were too similar

Refinements for next time: Add color palette guidance

SAMPLE ENTRY 3 — Technical Simplification

LEFT PAGE

PROJECT / CONTEXT:

Summarize a complex article into a simple explanation

ROLE / PERSONA:
Technical Simplification Writer

☑ Expert consultant
□ Creative writer
□ Data analyst
☑ Teacher / educator
□ Other: __________

TONE CHECKLIST:
☑ Professional
□ Casual / Friendly
□ Technical
☑ Educational
□ Creative

OUTPUT FORMAT:
□ Bullet points
□ Numbered list
☑ Paragraph (short)
□ Table / chart
□ Step by step

AUDIENCE:
Non technical adults who need straightforward understanding

KEY CONSTRAINTS:
Avoid jargon
Keep the summary within two short paragraphs

MEMO:
Use this prompt when simplifying research or technical content for general audiences.

SAMPLE ENTRY 3 — Technical Simplification

RIGHT PAGE

PROMPT TITLE: Simple Summary Prompt

DATE: 02 15 2025
VERSION: 1.0

TASK OR GOAL:
Turn complex content into an easy to understand summary

CONTEXT AND SPECIFICS:
Audience may not know technical concepts so keep explanations relatable and simple

EXAMPLES TO INCLUDE:
Good example:
Use everyday comparisons to explain complex ideas

What to avoid:
Avoid heavy terminology or multi step theories

STRUCTURE OR STEPS:

1. Identify main theme
2. Explain core idea
3. Add one relatable example
4. Limit to two paragraphs
5. No jargon
6. Keep tone neutral
7. Keep reading level simple
8. Maintain structure

ACTUAL PROMPT:
Act as a Technical Simplification Writer. Summarize the following content in two short paragraphs using plain English. Explain the core idea and include one everyday example. Avoid jargon and keep the summary easy to read.

RESULTS OR NOTES:
What worked: The two paragraph limit improved clarity

What didn't: Sometimes needed a clearer example

Refinements for next time: Specify reading level such as eighth grade

✦ SAMPLE ENTRY 4 — Business Communication

LEFT PAGE

PROJECT / CONTEXT:

Write a short customer outreach message

ROLE / PERSONA:

Professional Customer Engagement Specialist

☑ Expert consultant
□ Creative writer
□ Data analyst
□ Teacher / educator
□ Other: __________

TONE CHECKLIST:

☑ Professional
□ Casual / Friendly
□ Technical
☑ Educational/ (Warm)
□ Creative

OUTPUT FORMAT:

□ Bullet points
☑ Numbered List/ (Short email)
☑ Paragraph
□ Table / chart
□ Step by step

AUDIENCE:

Returning customers who have not engaged recently

KEY CONSTRAINTS:

Avoid sales heavy language
Keep message under one hundred twenty words

MEMO:

Use this for writing outreach messages follow ups or customer check ins.

SAMPLE ENTRY 4 — Business Communication

IGHT PAGE

ROMPT TITLE: Customer Check In Message

ATE: 02 15 2025

ERSION: 1.0

ASK OR GOAL:

/rite a friendly reconnection message to returning customers

ONTEXT AND SPECIFICS:

arget audience has purchased previously but has not engaged recently

XAMPLES TO INCLUDE:

ood example:

riendly tone expresses appreciation and provides a simple update

/hat to avoid:

void aggressive offers or pressure wording

TRUCTURE OR STEPS:

1. Warm greeting
2. Show appreciation
3. Offer a helpful update
4. Soft invitation to reconnect
5. Stay under one hundred twenty words
6. Maintain warm tone
7. Keep message focused
8. Keep instruction simple

CTUAL PROMPT:

ct as a Professional Customer Engagement Specialist. Write a warm friendly message nder one hundred twenty words to reconnect with returning customers. Include a brief hank you, one helpful update and a soft invitation to engage again. Keep tone sincere nd easy to read.

ESULTS OR NOTES:

/hat worked: Word limit created focus

/hat didn't: Some tone variations felt too formal

efinements for next time: Specify brand voice if needed

PROJECT/CONTEXT: ______________________

ROLE / PERSONA:______________________________________

□ Expert consultant
□ Creative writer
□ Data analyst
□ Teacher/educator
□ Other: __________________

TONE CHECKLIST:
□ Professional
□ Casual/friendly
□ Technical
□ Educational
□ Creative

OUTPUT FORMAT:
□ Bullet points
□ Numbered list
□ Paragraphs
□ Table/chart
□ Step-by-step

AUDIENCE: ____

KEY CONSTRAINTS:

____

____

____

MEMO ____

____

____

____

PROMPT TITLE: ______________________________

DATE: ______________________________ **VERSION:** ________________

TASK / GOAL:

________________________________ ________________________________

________________________________ ________________________________

CONTEXT & SPECIFICS:

________________________________ ________________________________

________________________________ ________________________________

EXAMPLES TO INCLUDE:

Good example:

________________________________ ________________________________

________________________________ ________________________________

What to avoid:

________________________________ ________________________________

________________________________ ________________________________

STRUCTURE/STEPS:

1. ______________________________ 5. ______________________________
2. ______________________________ 6. ______________________________
3. ______________________________ 7. ______________________________
4. ______________________________ 8. ______________________________

ACTUAL PROMPT:

____

____

____

RESULTS/NOTES:

What worked: ____

What didn't: ____

Refinements for next time:________________________________

____

PROJECT/CONTEXT: ______________________

ROLE / PERSONA:______________________________________

□ Expert consultant
□ Creative writer
□ Data analyst
□ Teacher/educator
□ Other: __________________

TONE CHECKLIST:
□ Professional
□ Casual/friendly
□ Technical
□ Educational
□ Creative

OUTPUT FORMAT:
□ Bullet points
□ Numbered list
□ Paragraphs
□ Table/chart
□ Step-by-step

AUDIENCE: ______________________________________

KEY CONSTRAINTS:

____

____

____

MEMO ____

____

____

____

PROMPT TITLE: ______________________

DATE: ______________ **VERSION:** __________

TASK / GOAL:

______________________ ______________________

______________________ ______________________

CONTEXT & SPECIFICS:

______________________ ______________________

______________________ ______________________

EXAMPLES TO INCLUDE:

Good example:

______________________ ______________________

______________________ ______________________

What to avoid:

______________________ ______________________

______________________ ______________________

STRUCTURE/STEPS:

1. ______________________
2. ______________________
3. ______________________
4. ______________________
5. ______________________
6. ______________________
7. ______________________
8. ______________________

ACTUAL PROMPT:

____

____

____

RESULTS/NOTES:

What worked: ______________________________

What didn't: ______________________________

Refinements for next time: ______________________

____

PROJECT/CONTEXT: ____________________

ROLE / PERSONA:______________________________

- ☐ Expert consultant
- ☐ Creative writer
- ☐ Data analyst
- ☐ Teacher/educator
- ☐ Other: ______________

TONE CHECKLIST:

- ☐ Professional
- ☐ Casual/friendly
- ☐ Technical
- ☐ Educational
- ☐ Creative

OUTPUT FORMAT:

- ☐ Bullet points
- ☐ Numbered list
- ☐ Paragraphs
- ☐ Table/chart
- ☐ Step-by-step

AUDIENCE: ______________________________

KEY CONSTRAINTS:

____

____

____

MEMO ______________________________

____

____

____

PROMPT TITLE: ____________________________

DATE: ______________________ **VERSION:** ____________

TASK / GOAL:

______________________ ______________________

______________________ ______________________

CONTEXT & SPECIFICS:

______________________ ______________________

______________________ ______________________

EXAMPLES TO INCLUDE:

Good example:

______________________ ______________________

______________________ ______________________

What to avoid:

______________________ ______________________

______________________ ______________________

STRUCTURE/STEPS:

1. ______________________ 5. ______________________

2. ______________________ 6. ______________________

3. ______________________ 7. ______________________

4. ______________________ 8. ______________________

ACTUAL PROMPT:

____

____

____

RESULTS/NOTES:

What worked: ________________________________

What didn't: ________________________________

Refinements for next time: ____________________

____

PROJECT/CONTEXT: ________________________

ROLE / PERSONA:____

□ Expert consultant
□ Creative writer
□ Data analyst
□ Teacher/educator
□ Other: ____________________

TONE CHECKLIST:
□ Professional
□ Casual/friendly
□ Technical
□ Educational
□ Creative

OUTPUT FORMAT:
□ Bullet points
□ Numbered list
□ Paragraphs
□ Table/chart
□ Step-by-step

AUDIENCE: ____

KEY CONSTRAINTS:

____

____

____

MEMO ____

____

____

____

PROMPT TITLE: ______________________________

DATE: ______________________ **VERSION:** _____________

TASK / GOAL:

______________________ ______________________

______________________ ______________________

CONTEXT & SPECIFICS:

______________________ ______________________

______________________ ______________________

EXAMPLES TO INCLUDE:

Good example:

______________________ ______________________

______________________ ______________________

What to avoid:

______________________ ______________________

______________________ ______________________

STRUCTURE/STEPS:

1. ______________________
2. ______________________
3. ______________________
4. ______________________
5. ______________________
6. ______________________
7. ______________________
8. ______________________

ACTUAL PROMPT:

____

____

____

RESULTS/NOTES:

What worked: ______________________________

What didn't: ______________________________

Refinements for next time: ______________________

____

PROJECT/CONTEXT: ____________________

ROLE / PERSONA:______________________________

□ Expert consultant
□ Creative writer
□ Data analyst
□ Teacher/educator
□ Other: ______________

TONE CHECKLIST:
□ Professional
□ Casual/friendly
□ Technical
□ Educational
□ Creative

OUTPUT FORMAT:
□ Bullet points
□ Numbered list
□ Paragraphs
□ Table/chart
□ Step-by-step

AUDIENCE: ______________________________

KEY CONSTRAINTS:

____

____

____

MEMO ______________________________

____

____

____

PROMPT TITLE: ______________________________

DATE: ______________________ **VERSION:** ____________

TASK / GOAL:

________________________ ________________________

________________________ ________________________

CONTEXT & SPECIFICS:

________________________ ________________________

________________________ ________________________

EXAMPLES TO INCLUDE:

Good example:

________________________ ________________________

________________________ ________________________

What to avoid:

________________________ ________________________

________________________ ________________________

STRUCTURE/STEPS:

1. ______________________ 5.______________________

2. ______________________ 6.______________________

3. ______________________ 7.______________________

4. ______________________ 8.______________________

ACTUAL PROMPT:

____

____

____

RESULTS/NOTES:

What worked: ______________________________________

What didn't: ______________________________________

Refinements for next time:______________________________

____

PROJECT/CONTEXT: ______________________

ROLE / PERSONA:______________________________

- □ Expert consultant
- □ Creative writer
- □ Data analyst
- □ Teacher/educator
- □ Other: ______________

TONE CHECKLIST:

- □ Professional
- □ Casual/friendly
- □ Technical
- □ Educational
- □ Creative

OUTPUT FORMAT:

- □ Bullet points
- □ Numbered list
- □ Paragraphs
- □ Table/chart
- □ Step-by-step

AUDIENCE: ______________________________

KEY CONSTRAINTS:

____

____

____

MEMO ______________________________

____

____

____

PROMPT TITLE: ______________________________

DATE: ______________________ **VERSION:** ______________

TASK / GOAL:

______________________ ______________________

______________________ ______________________

CONTEXT & SPECIFICS:

______________________ ______________________

______________________ ______________________

EXAMPLES TO INCLUDE:

Good example:

______________________ ______________________

______________________ ______________________

What to avoid:

______________________ ______________________

______________________ ______________________

STRUCTURE/STEPS:

1. ______________________
2. ______________________
3. ______________________
4. ______________________
5. ______________________
6. ______________________
7. ______________________
8. ______________________

ACTUAL PROMPT:

____

____

____

RESULTS/NOTES:

What worked: ______________________________

What didn't: ______________________________

Refinements for next time: ______________________

____

PROJECT/CONTEXT: ____________________

ROLE / PERSONA:________________________________

□ Expert consultant
□ Creative writer
□ Data analyst
□ Teacher/educator
□ Other: ________________

TONE CHECKLIST:
□ Professional
□ Casual/friendly
□ Technical
□ Educational
□ Creative

OUTPUT FORMAT:
□ Bullet points
□ Numbered list
□ Paragraphs
□ Table/chart
□ Step-by-step

AUDIENCE: ____

KEY CONSTRAINTS:

____

____

____

MEMO ____

____

____

____

PROMPT TITLE: ______________________________

DATE: ______________________ **VERSION:** ____________

TASK / GOAL:

________________________ ________________________

________________________ ________________________

CONTEXT & SPECIFICS:

________________________ ________________________

________________________ ________________________

EXAMPLES TO INCLUDE:

Good example:

________________________ ________________________

________________________ ________________________

What to avoid:

________________________ ________________________

________________________ ________________________

STRUCTURE/STEPS:

1. ______________________ 5. ______________________

2. ______________________ 6. ______________________

3. ______________________ 7. ______________________

4. ______________________ 8. ______________________

ACTUAL PROMPT:

____

____

____

RESULTS/NOTES:

What worked: ______________________________________

What didn't: ______________________________________

Refinements for next time: ___________________________

____

PROJECT/CONTEXT: ____________________

ROLE / PERSONA:____________________________________

□ Expert consultant
□ Creative writer
□ Data analyst
□ Teacher/educator
□ Other: __________________

TONE CHECKLIST:
□ Professional
□ Casual/friendly
□ Technical
□ Educational
□ Creative

OUTPUT FORMAT:
□ Bullet points
□ Numbered list
□ Paragraphs
□ Table/chart
□ Step-by-step

AUDIENCE: ____

KEY CONSTRAINTS:

____

____

____

MEMO ____

____

____

____

PROMPT TITLE: ____________________

DATE: ____________________ **VERSION:** ____________

TASK / GOAL:

____________________ ____________________

____________________ ____________________

CONTEXT & SPECIFICS:

____________________ ____________________

____________________ ____________________

EXAMPLES TO INCLUDE:

Good example:

____________________ ____________________

____________________ ____________________

What to avoid:

____________________ ____________________

____________________ ____________________

STRUCTURE/STEPS:

1. ____________________
2. ____________________
3. ____________________
4. ____________________
5. ____________________
6. ____________________
7. ____________________
8. ____________________

ACTUAL PROMPT:

____

____

____

RESULTS/NOTES:

What worked: ____________________

What didn't: ____________________

Refinements for next time: ____________________

____

PROJECT/CONTEXT: ____________________

ROLE / PERSONA:____________________________________

- □ Expert consultant
- □ Creative writer
- □ Data analyst
- □ Teacher/educator
- □ Other: ________________

TONE CHECKLIST:

- □ Professional
- □ Casual/friendly
- □ Technical
- □ Educational
- □ Creative

OUTPUT FORMAT:

- □ Bullet points
- □ Numbered list
- □ Paragraphs
- □ Table/chart
- □ Step-by-step

AUDIENCE: ____

KEY CONSTRAINTS:

____

____

____

MEMO ____

____

____

____

PROMPT TITLE: ______________________________

DATE: ____________________ **VERSION:** ______________

TASK / GOAL:

______________________ ______________________

______________________ ______________________

CONTEXT & SPECIFICS:

______________________ ______________________

______________________ ______________________

EXAMPLES TO INCLUDE:

Good example:

______________________ ______________________

______________________ ______________________

What to avoid:

______________________ ______________________

______________________ ______________________

STRUCTURE/STEPS:

1. ______________________
2. ______________________
3. ______________________
4. ______________________
5. ______________________
6. ______________________
7. ______________________
8. ______________________

ACTUAL PROMPT:

____

____

____

RESULTS/NOTES:

What worked: ______________________________

What didn't: ______________________________

Refinements for next time: ______________________

____

PROJECT/CONTEXT: ______________________

ROLE / PERSONA:______________________________________

□ Expert consultant
□ Creative writer
□ Data analyst
□ Teacher/educator
□ Other: ____________________

TONE CHECKLIST:
□ Professional
□ Casual/friendly
□ Technical
□ Educational
□ Creative

OUTPUT FORMAT:
□ Bullet points
□ Numbered list
□ Paragraphs
□ Table/chart
□ Step-by-step

AUDIENCE: ____

KEY CONSTRAINTS:

____

____

____

MEMO ____

____

____

____

PROMPT TITLE: ____________________

DATE: ____________________ **VERSION:** ____________

TASK / GOAL:

____________________ ____________________

____________________ ____________________

CONTEXT & SPECIFICS:

____________________ ____________________

____________________ ____________________

EXAMPLES TO INCLUDE:

Good example:

____________________ ____________________

____________________ ____________________

What to avoid:

____________________ ____________________

____________________ ____________________

STRUCTURE/STEPS:

1. ____________________
2. ____________________
3. ____________________
4. ____________________
5. ____________________
6. ____________________
7. ____________________
8. ____________________

ACTUAL PROMPT:

____

____

____

RESULTS/NOTES:

What worked: ____________________

What didn't: ____________________

Refinements for next time: ____________________

____

PROJECT/CONTEXT: ____________________

ROLE / PERSONA:________________________________

□ Expert consultant
□ Creative writer
□ Data analyst
□ Teacher/educator
□ Other: ____________________

TONE CHECKLIST:
□ Professional
□ Casual/friendly
□ Technical
□ Educational
□ Creative

OUTPUT FORMAT:
□ Bullet points
□ Numbered list
□ Paragraphs
□ Table/chart
□ Step-by-step

AUDIENCE: ________________________________

KEY CONSTRAINTS:

____

____

____

MEMO ________________________________

____

____

____

PROMPT TITLE: ____________________

DATE: ____________________ **VERSION:** ____________

TASK / GOAL:

____________________ ____________________

____________________ ____________________

CONTEXT & SPECIFICS:

____________________ ____________________

____________________ ____________________

EXAMPLES TO INCLUDE:

Good example:

____________________ ____________________

____________________ ____________________

What to avoid:

____________________ ____________________

____________________ ____________________

STRUCTURE/STEPS:

1. ____________________
2. ____________________
3. ____________________
4. ____________________
5. ____________________
6. ____________________
7. ____________________
8. ____________________

ACTUAL PROMPT:

____

____

____

RESULTS/NOTES:

What worked: ____________________

What didn't: ____________________

Refinements for next time: ____________________

____

PROJECT/CONTEXT: ______________________

ROLE / PERSONA: ______________________________________

□ Expert consultant
□ Creative writer
□ Data analyst
□ Teacher/educator
□ Other: ____________________

TONE CHECKLIST:
□ Professional
□ Casual/friendly
□ Technical
□ Educational
□ Creative

OUTPUT FORMAT:
□ Bullet points
□ Numbered list
□ Paragraphs
□ Table/chart
□ Step-by-step

AUDIENCE: ____

KEY CONSTRAINTS:

____

____

____

MEMO ____

____

____

____

PROMPT TITLE: ______________________________

DATE: ______________________ **VERSION:** ____________

TASK / GOAL:

________________________ ______________________

________________________ ______________________

CONTEXT & SPECIFICS:

________________________ ______________________

________________________ ______________________

EXAMPLES TO INCLUDE:

Good example:

________________________ ______________________

________________________ ______________________

What to avoid:

________________________ ______________________

________________________ ______________________

STRUCTURE/STEPS:

1. ______________________ 5. ______________________
2. ______________________ 6. ______________________
3. ______________________ 7. ______________________
4. ______________________ 8. ______________________

ACTUAL PROMPT:

____

____

____

RESULTS/NOTES:

What worked: ______________________________

What didn't: ______________________________

Refinements for next time:______________________

____

PROJECT/CONTEXT: ____________________

ROLE / PERSONA:____________________________

□ Expert consultant
□ Creative writer
□ Data analyst
□ Teacher/educator
□ Other: ______________

TONE CHECKLIST:
□ Professional
□ Casual/friendly
□ Technical
□ Educational
□ Creative

OUTPUT FORMAT:
□ Bullet points
□ Numbered list
□ Paragraphs
□ Table/chart
□ Step-by-step

AUDIENCE: ____________________________

KEY CONSTRAINTS:

____

____

____

MEMO ____________________________________

____

____

____

PROMPT TITLE: ______________________________

DATE: ______________________ **VERSION:** _____________

TASK / GOAL:

________________________ ________________________

________________________ ________________________

CONTEXT & SPECIFICS:

________________________ ________________________

________________________ ________________________

EXAMPLES TO INCLUDE:

Good example:

________________________ ________________________

________________________ ________________________

What to avoid:

________________________ ________________________

________________________ ________________________

STRUCTURE/STEPS:

1. ______________________ 5. ______________________

2. ______________________ 6. ______________________

3. ______________________ 7. ______________________

4. ______________________ 8. ______________________

ACTUAL PROMPT:

____

____

____

RESULTS/NOTES:

What worked: ______________________________________

What didn't: ______________________________________

Refinements for next time:______________________________

____

PROJECT/CONTEXT: ____________________

ROLE / PERSONA:____

□ Expert consultant
□ Creative writer
□ Data analyst
□ Teacher/educator
□ Other: ____________________

TONE CHECKLIST:
□ Professional
□ Casual/friendly
□ Technical
□ Educational
□ Creative

OUTPUT FORMAT:
□ Bullet points
□ Numbered list
□ Paragraphs
□ Table/chart
□ Step-by-step

AUDIENCE: ____

KEY CONSTRAINTS:

____

____

____

MEMO ____

____

____

____

PROMPT TITLE: ______________________________

DATE: ______________________ **VERSION:** _____________

TASK / GOAL:

______________________________ ______________________________

______________________________ ______________________________

CONTEXT & SPECIFICS:

______________________________ ______________________________

______________________________ ______________________________

EXAMPLES TO INCLUDE:

Good example:

______________________________ ______________________________

______________________________ ______________________________

What to avoid:

______________________________ ______________________________

______________________________ ______________________________

STRUCTURE/STEPS:

1. ____________________ 5. ____________________

2. ____________________ 6. ____________________

3. ____________________ 7. ____________________

4. ____________________ 8. ____________________

ACTUAL PROMPT:

____

____

____

RESULTS/NOTES:

What worked: ______________________________________

What didn't: ______________________________________

Refinements for next time: ____________________________

____

PROJECT/CONTEXT: ________________________

ROLE / PERSONA:_________________________________

□ Expert consultant
□ Creative writer
□ Data analyst
□ Teacher/educator
□ Other: ________________

TONE CHECKLIST:
□ Professional
□ Casual/friendly
□ Technical
□ Educational
□ Creative

OUTPUT FORMAT:
□ Bullet points
□ Numbered list
□ Paragraphs
□ Table/chart
□ Step-by-step

AUDIENCE: ____

KEY CONSTRAINTS:

____

____

____

MEMO ____

____

____

____

PROMPT TITLE: ______________________________

DATE: ______________________ **VERSION:** ______________

TASK / GOAL:

________________________ ________________________

________________________ ________________________

CONTEXT & SPECIFICS:

________________________ ________________________

________________________ ________________________

EXAMPLES TO INCLUDE:

Good example:

________________________ ________________________

________________________ ________________________

What to avoid:

________________________ ________________________

________________________ ________________________

STRUCTURE/STEPS:

1. ______________________ 5. ______________________

2. ______________________ 6. ______________________

3. ______________________ 7. ______________________

4. ______________________ 8. ______________________

ACTUAL PROMPT:

____

____

____

RESULTS/NOTES:

What worked: ______________________________________

What didn't: ______________________________________

Refinements for next time: __________________________

____

PROJECT/CONTEXT: ______________________

ROLE / PERSONA:______________________________________

□ Expert consultant
□ Creative writer
□ Data analyst
□ Teacher/educator
□ Other: __________________

TONE CHECKLIST:
□ Professional
□ Casual/friendly
□ Technical
□ Educational
□ Creative

OUTPUT FORMAT:
□ Bullet points
□ Numbered list
□ Paragraphs
□ Table/chart
□ Step-by-step

AUDIENCE: ____

KEY CONSTRAINTS:

____

____

____

MEMO ____

____

____

____

PROMPT TITLE: ______________________________

DATE: ______________________________ **VERSION:** ______________

TASK / GOAL:

______________________________ ______________________________

______________________________ ______________________________

CONTEXT & SPECIFICS:

______________________________ ______________________________

______________________________ ______________________________

EXAMPLES TO INCLUDE:

Good example:

______________________________ ______________________________

______________________________ ______________________________

What to avoid:

______________________________ ______________________________

______________________________ ______________________________

STRUCTURE/STEPS:

1. ______________________________
2. ______________________________
3. ______________________________
4. ______________________________
5. ______________________________
6. ______________________________
7. ______________________________
8. ______________________________

ACTUAL PROMPT:

____

____

____

RESULTS/NOTES:

What worked: ____

What didn't: ____

Refinements for next time: ______________________________________

____

PROJECT/CONTEXT: ______________________

ROLE / PERSONA: ______________________________

□ Expert consultant
□ Creative writer
□ Data analyst
□ Teacher/educator
□ Other: ______________

TONE CHECKLIST:
□ Professional
□ Casual/friendly
□ Technical
□ Educational
□ Creative

OUTPUT FORMAT:
□ Bullet points
□ Numbered list
□ Paragraphs
□ Table/chart
□ Step-by-step

AUDIENCE: ____________________________________

KEY CONSTRAINTS:

____

____

____

MEMO ______________________________________

____

____

____

PROMPT TITLE: ______________________________

DATE: ______________________ **VERSION:** ____________

TASK / GOAL:

________________________ ________________________

________________________ ________________________

CONTEXT & SPECIFICS:

________________________ ________________________

________________________ ________________________

EXAMPLES TO INCLUDE:

Good example:

________________________ ________________________

________________________ ________________________

What to avoid:

________________________ ________________________

________________________ ________________________

STRUCTURE/STEPS:

1. ______________________ 5. ______________________

2. ______________________ 6. ______________________

3. ______________________ 7. ______________________

4. ______________________ 8. ______________________

ACTUAL PROMPT:

____

____

____

RESULTS/NOTES:

What worked: ______________________________________

What didn't: ______________________________________

Refinements for next time: ____________________________

____

PROJECT/CONTEXT: ______________________

ROLE / PERSONA:______________________________________

□ Expert consultant
□ Creative writer
□ Data analyst
□ Teacher/educator
□ Other: __________________

TONE CHECKLIST:
□ Professional
□ Casual/friendly
□ Technical
□ Educational
□ Creative

OUTPUT FORMAT:
□ Bullet points
□ Numbered list
□ Paragraphs
□ Table/chart
□ Step-by-step

AUDIENCE: ____

KEY CONSTRAINTS:

____

____

____

MEMO ____

____

____

____

PROMPT TITLE: ______________________________

DATE: ______________________ **VERSION:** ______________

TASK / GOAL:

______________________ ______________________

______________________ ______________________

CONTEXT & SPECIFICS:

______________________ ______________________

______________________ ______________________

EXAMPLES TO INCLUDE:

Good example:

______________________ ______________________

______________________ ______________________

What to avoid:

______________________ ______________________

______________________ ______________________

STRUCTURE/STEPS:

1. ______________________ 5. ______________________

2. ______________________ 6. ______________________

3. ______________________ 7. ______________________

4. ______________________ 8. ______________________

ACTUAL PROMPT:

____

____

____

RESULTS/NOTES:

What worked: ______________________________

What didn't: ______________________________

Refinements for next time: ______________________

____

PROJECT/CONTEXT: ______________________

ROLE / PERSONA:______________________________

□ Expert consultant
□ Creative writer
□ Data analyst
□ Teacher/educator
□ Other: ________________

TONE CHECKLIST:
□ Professional
□ Casual/friendly
□ Technical
□ Educational
□ Creative

OUTPUT FORMAT:
□ Bullet points
□ Numbered list
□ Paragraphs
□ Table/chart
□ Step-by-step

AUDIENCE: ____________________________________

KEY CONSTRAINTS:

____

____

____

MEMO ____________________________________

____

____

____

PROMPT TITLE: ______________________________

DATE: ______________________ **VERSION:** ____________

TASK / GOAL:

______________________ ______________________

______________________ ______________________

CONTEXT & SPECIFICS:

______________________ ______________________

______________________ ______________________

EXAMPLES TO INCLUDE:

Good example:

______________________ ______________________

______________________ ______________________

What to avoid:

______________________ ______________________

______________________ ______________________

STRUCTURE/STEPS:

1. ______________________
2. ______________________
3. ______________________
4. ______________________
5. ______________________
6. ______________________
7. ______________________
8. ______________________

ACTUAL PROMPT:

____

____

____

RESULTS/NOTES:

What worked: ______________________________

What didn't: ______________________________

Refinements for next time: ______________________

____

PROJECT/CONTEXT: ____________________

ROLE / PERSONA:______________________________

- □ Expert consultant
- □ Creative writer
- □ Data analyst
- □ Teacher/educator
- □ Other: ______________

TONE CHECKLIST:

- □ Professional
- □ Casual/friendly
- □ Technical
- □ Educational
- □ Creative

OUTPUT FORMAT:

- □ Bullet points
- □ Numbered list
- □ Paragraphs
- □ Table/chart
- □ Step-by-step

AUDIENCE: ___________________________________

KEY CONSTRAINTS:

____

____

____

MEMO ______________________________________

____

____

____

PROMPT TITLE: ____________________________

DATE: ______________________ **VERSION:** ______________

TASK / GOAL:

________________________ ________________________

________________________ ________________________

CONTEXT & SPECIFICS:

________________________ ________________________

________________________ ________________________

EXAMPLES TO INCLUDE:

Good example:

________________________ ________________________

________________________ ________________________

What to avoid:

________________________ ________________________

________________________ ________________________

STRUCTURE/STEPS:

1. ____________________ 5. ____________________

2. ____________________ 6. ____________________

3. ____________________ 7. ____________________

4. ____________________ 8. ____________________

ACTUAL PROMPT:

____

____

____

RESULTS/NOTES:

What worked: ______________________________________

What didn't: ______________________________________

Refinements for next time:____________________________

____

PROJECT/CONTEXT: ________________________

ROLE / PERSONA:____

□ Expert consultant
□ Creative writer
□ Data analyst
□ Teacher/educator
□ Other: ____________________

TONE CHECKLIST:
□ Professional
□ Casual/friendly
□ Technical
□ Educational
□ Creative

OUTPUT FORMAT:
□ Bullet points
□ Numbered list
□ Paragraphs
□ Table/chart
□ Step-by-step

AUDIENCE: ____

KEY CONSTRAINTS:

____

____

____

MEMO ____

____

____

____

PROMPT TITLE: ______________________________

DATE: ______________________ **VERSION:** ____________

TASK / GOAL:

________________________ ________________________

________________________ ________________________

CONTEXT & SPECIFICS:

________________________ ________________________

________________________ ________________________

EXAMPLES TO INCLUDE:

Good example:

________________________ ________________________

________________________ ________________________

What to avoid:

________________________ ________________________

________________________ ________________________

STRUCTURE/STEPS:

1. ______________________ 5. ______________________

2. ______________________ 6. ______________________

3. ______________________ 7. ______________________

4. ______________________ 8. ______________________

ACTUAL PROMPT:

____

____

____

RESULTS/NOTES:

What worked: ______________________________________

What didn't: ______________________________________

Refinements for next time: ____________________________

____

PROJECT/CONTEXT: ______________________

ROLE / PERSONA:______________________________________

□ Expert consultant
□ Creative writer
□ Data analyst
□ Teacher/educator
□ Other: ___________________

TONE CHECKLIST:
□ Professional
□ Casual/friendly
□ Technical
□ Educational
□ Creative

OUTPUT FORMAT:
□ Bullet points
□ Numbered list
□ Paragraphs
□ Table/chart
□ Step-by-step

AUDIENCE: ____

KEY CONSTRAINTS:

____

____

____

MEMO ____

____

____

____

PROMPT TITLE: ______________________

DATE: ____________ **VERSION:** ____________

TASK / GOAL:

______________________ ______________________

______________________ ______________________

CONTEXT & SPECIFICS:

______________________ ______________________

______________________ ______________________

EXAMPLES TO INCLUDE:

Good example:

______________________ ______________________

______________________ ______________________

What to avoid:

______________________ ______________________

______________________ ______________________

STRUCTURE/STEPS:

1. ______________________
2. ______________________
3. ______________________
4. ______________________
5. ______________________
6. ______________________
7. ______________________
8. ______________________

ACTUAL PROMPT:

____

____

____

RESULTS/NOTES:

What worked: ______________________________

What didn't: ______________________________

Refinements for next time: ______________________

____

PROJECT/CONTEXT: ____________________

ROLE / PERSONA:____________________________________

□ Expert consultant
□ Creative writer
□ Data analyst
□ Teacher/educator
□ Other: __________________

TONE CHECKLIST:
□ Professional
□ Casual/friendly
□ Technical
□ Educational
□ Creative

OUTPUT FORMAT:
□ Bullet points
□ Numbered list
□ Paragraphs
□ Table/chart
□ Step-by-step

AUDIENCE: ____

KEY CONSTRAINTS:

____

____

____

MEMO ____

____

____

____

PROMPT TITLE: ______________________________

DATE: ____________________ **VERSION:** ____________

TASK / GOAL:

______________________ ______________________

______________________ ______________________

CONTEXT & SPECIFICS:

______________________ ______________________

______________________ ______________________

EXAMPLES TO INCLUDE:

Good example:

______________________ ______________________

______________________ ______________________

What to avoid:

______________________ ______________________

______________________ ______________________

STRUCTURE/STEPS:

1. ____________________
2. ____________________
3. ____________________
4. ____________________
5. ____________________
6. ____________________
7. ____________________
8. ____________________

ACTUAL PROMPT:

____

____

____

RESULTS/NOTES:

What worked: ______________________________

What didn't: ______________________________

Refinements for next time:______________________

____

PROJECT/CONTEXT: ______________________

ROLE / PERSONA:______________________________

□ Expert consultant
□ Creative writer
□ Data analyst
□ Teacher/educator
□ Other: ______________

TONE CHECKLIST:
□ Professional
□ Casual/friendly
□ Technical
□ Educational
□ Creative

OUTPUT FORMAT:
□ Bullet points
□ Numbered list
□ Paragraphs
□ Table/chart
□ Step-by-step

AUDIENCE: ___________________________________

KEY CONSTRAINTS:

____

____

____

MEMO ______________________________________

____

____

____

PROMPT TITLE: ______________________________

DATE: ______________________________ **VERSION:** ________________

TASK / GOAL:

______________________________ ______________________________

______________________________ ______________________________

CONTEXT & SPECIFICS:

______________________________ ______________________________

______________________________ ______________________________

EXAMPLES TO INCLUDE:

Good example:

______________________________ ______________________________

______________________________ ______________________________

What to avoid:

______________________________ ______________________________

______________________________ ______________________________

STRUCTURE/STEPS:

1. ______________________________
2. ______________________________
3. ______________________________
4. ______________________________
5. ______________________________
6. ______________________________
7. ______________________________
8. ______________________________

ACTUAL PROMPT:

____

____

____

RESULTS/NOTES:

What worked: ____

What didn't: ____

Refinements for next time: ________________________________

____

PROJECT/CONTEXT: ______________________

ROLE / PERSONA:______________________________________

□ Expert consultant
□ Creative writer
□ Data analyst
□ Teacher/educator
□ Other: ___________________

TONE CHECKLIST:
□ Professional
□ Casual/friendly
□ Technical
□ Educational
□ Creative

OUTPUT FORMAT:
□ Bullet points
□ Numbered list
□ Paragraphs
□ Table/chart
□ Step-by-step

AUDIENCE: ____

KEY CONSTRAINTS:

____

____

____

MEMO ____

____

____

____

PROMPT TITLE: ____________________

DATE: ____________________ **VERSION:** ____________

TASK / GOAL:

____________________ ____________________

____________________ ____________________

CONTEXT & SPECIFICS:

____________________ ____________________

____________________ ____________________

EXAMPLES TO INCLUDE:

Good example:

____________________ ____________________

____________________ ____________________

What to avoid:

____________________ ____________________

____________________ ____________________

STRUCTURE/STEPS:

1. ____________________
2. ____________________
3. ____________________
4. ____________________
5. ____________________
6. ____________________
7. ____________________
8. ____________________

ACTUAL PROMPT:

____

____

____

RESULTS/NOTES:

What worked: ____________________

What didn't: ____________________

Refinements for next time: ____________________

____

PROJECT/CONTEXT: ______________________

ROLE / PERSONA:______________________________

□ Expert consultant
□ Creative writer
□ Data analyst
□ Teacher/educator
□ Other: ______________

TONE CHECKLIST:
□ Professional
□ Casual/friendly
□ Technical
□ Educational
□ Creative

OUTPUT FORMAT:
□ Bullet points
□ Numbered list
□ Paragraphs
□ Table/chart
□ Step-by-step

AUDIENCE: ____________________________________

KEY CONSTRAINTS:

____

____

____

MEMO ______________________________________

____

____

____

PROMPT TITLE: ______________________

DATE: ______________________ **VERSION:** ______________

TASK / GOAL:

______________________ ______________________

______________________ ______________________

CONTEXT & SPECIFICS:

______________________ ______________________

______________________ ______________________

EXAMPLES TO INCLUDE:

Good example:

______________________ ______________________

______________________ ______________________

What to avoid:

______________________ ______________________

______________________ ______________________

STRUCTURE/STEPS:

1. ______________________
2. ______________________
3. ______________________
4. ______________________
5. ______________________
6. ______________________
7. ______________________
8. ______________________

ACTUAL PROMPT:

____

____

____

RESULTS/NOTES:

What worked: ______________________________

What didn't: ______________________________

Refinements for next time: ______________________

____

PROJECT/CONTEXT: ______________________

ROLE / PERSONA:______________________________________

□ Expert consultant
□ Creative writer
□ Data analyst
□ Teacher/educator
□ Other: ___________________

TONE CHECKLIST:
□ Professional
□ Casual/friendly
□ Technical
□ Educational
□ Creative

OUTPUT FORMAT:
□ Bullet points
□ Numbered list
□ Paragraphs
□ Table/chart
□ Step-by-step

AUDIENCE: ____

KEY CONSTRAINTS:

____

____

____

MEMO ____

____

____

____

PROMPT TITLE: ______________________________

DATE: ______________________ **VERSION:** _____________

TASK / GOAL:

________________________ ________________________

________________________ ________________________

CONTEXT & SPECIFICS:

________________________ ________________________

________________________ ________________________

EXAMPLES TO INCLUDE:

Good example:

________________________ ________________________

________________________ ________________________

What to avoid:

________________________ ________________________

________________________ ________________________

STRUCTURE/STEPS:

1. ______________________
2. ______________________
3. ______________________
4. ______________________
5. ______________________
6. ______________________
7. ______________________
8. ______________________

ACTUAL PROMPT:

____

____

____

RESULTS/NOTES:

What worked: ______________________________________

What didn't: ______________________________________

Refinements for next time: ____________________________

____

PROJECT/CONTEXT: ______________________

ROLE / PERSONA:______________________________

□ Expert consultant
□ Creative writer
□ Data analyst
□ Teacher/educator
□ Other: ______________

TONE CHECKLIST:
□ Professional
□ Casual/friendly
□ Technical
□ Educational
□ Creative

OUTPUT FORMAT:
□ Bullet points
□ Numbered list
□ Paragraphs
□ Table/chart
□ Step-by-step

AUDIENCE: ____________________________________

KEY CONSTRAINTS:

____

____

____

MEMO ______________________________________

____

____

____

PROMPT TITLE: ______________________________

DATE: ______________________ **VERSION:** ______________

TASK / GOAL:

______________________________ ______________________________

______________________________ ______________________________

CONTEXT & SPECIFICS:

______________________________ ______________________________

______________________________ ______________________________

EXAMPLES TO INCLUDE:

Good example:

______________________________ ______________________________

______________________________ ______________________________

What to avoid:

______________________________ ______________________________

______________________________ ______________________________

STRUCTURE/STEPS:

1. ______________________________
2. ______________________________
3. ______________________________
4. ______________________________
5. ______________________________
6. ______________________________
7. ______________________________
8. ______________________________

ACTUAL PROMPT:

____

____

____

RESULTS/NOTES:

What worked: ____

What didn't: ____

Refinements for next time: ________________________________

____

PROJECT/CONTEXT: ______________________

ROLE / PERSONA:____

□ Expert consultant
□ Creative writer
□ Data analyst
□ Teacher/educator
□ Other: ____________________

TONE CHECKLIST:
□ Professional
□ Casual/friendly
□ Technical
□ Educational
□ Creative

OUTPUT FORMAT:
□ Bullet points
□ Numbered list
□ Paragraphs
□ Table/chart
□ Step-by-step

AUDIENCE: ____

KEY CONSTRAINTS:

____

____

____

MEMO ____

____

____

____

PROMPT TITLE: ______________________

DATE: ______________________ **VERSION:** ______________

TASK / GOAL:

______________________ ______________________

______________________ ______________________

CONTEXT & SPECIFICS:

______________________ ______________________

______________________ ______________________

EXAMPLES TO INCLUDE:

Good example:

______________________ ______________________

______________________ ______________________

What to avoid:

______________________ ______________________

______________________ ______________________

STRUCTURE/STEPS:

1. ______________________ 5. ______________________

2. ______________________ 6. ______________________

3. ______________________ 7. ______________________

4. ______________________ 8. ______________________

ACTUAL PROMPT:

____

____

____

RESULTS/NOTES:

What worked: ______________________________

What didn't: ______________________________

Refinements for next time: ______________________

____

PROJECT/CONTEXT: ______________________

ROLE / PERSONA:________________________________

- □ Expert consultant
- □ Creative writer
- □ Data analyst
- □ Teacher/educator
- □ Other: ________________

TONE CHECKLIST:

- □ Professional
- □ Casual/friendly
- □ Technical
- □ Educational
- □ Creative

OUTPUT FORMAT:

- □ Bullet points
- □ Numbered list
- □ Paragraphs
- □ Table/chart
- □ Step-by-step

AUDIENCE: ____________________________________

KEY CONSTRAINTS:

____

____

____

MEMO ____________________________________

____

____

____

PROMPT TITLE: ____________________________

DATE: ______________________ **VERSION:** ______________

TASK / GOAL:

______________________ ______________________

______________________ ______________________

CONTEXT & SPECIFICS:

______________________ ______________________

______________________ ______________________

EXAMPLES TO INCLUDE:

Good example:

______________________ ______________________

______________________ ______________________

What to avoid:

______________________ ______________________

______________________ ______________________

STRUCTURE/STEPS:

1. ______________________
2. ______________________
3. ______________________
4. ______________________
5. ______________________
6. ______________________
7. ______________________
8. ______________________

ACTUAL PROMPT:

____

____

____

RESULTS/NOTES:

What worked: ______________________________

What didn't: ______________________________

Refinements for next time: ______________________

____

PROJECT/CONTEXT: ____________________

ROLE / PERSONA:______________________________

□ Expert consultant
□ Creative writer
□ Data analyst
□ Teacher/educator
□ Other: ____________________

TONE CHECKLIST:
□ Professional
□ Casual/friendly
□ Technical
□ Educational
□ Creative

OUTPUT FORMAT:
□ Bullet points
□ Numbered list
□ Paragraphs
□ Table/chart
□ Step-by-step

AUDIENCE: ______________________________

KEY CONSTRAINTS:

____

____

____

MEMO ______________________________

____

____

____

PROMPT TITLE: ______________________

DATE: ______________________ **VERSION:** ______________

TASK / GOAL:

______________________ ______________________

______________________ ______________________

CONTEXT & SPECIFICS:

______________________ ______________________

______________________ ______________________

EXAMPLES TO INCLUDE:

Good example:

______________________ ______________________

______________________ ______________________

What to avoid:

______________________ ______________________

______________________ ______________________

STRUCTURE/STEPS:

1. ______________________ 5. ______________________
2. ______________________ 6. ______________________
3. ______________________ 7. ______________________
4. ______________________ 8. ______________________

ACTUAL PROMPT:

____

____

____

RESULTS/NOTES:

What worked: ______________________________

What didn't: ______________________________

Refinements for next time: ______________________

____

PROJECT/CONTEXT: ____________________

ROLE / PERSONA:________________________________

- □ Expert consultant
- □ Creative writer
- □ Data analyst
- □ Teacher/educator
- □ Other: ____________________

TONE CHECKLIST:

- □ Professional
- □ Casual/friendly
- □ Technical
- □ Educational
- □ Creative

OUTPUT FORMAT:

- □ Bullet points
- □ Numbered list
- □ Paragraphs
- □ Table/chart
- □ Step-by-step

AUDIENCE: ________________________________

KEY CONSTRAINTS:

____

____

____

MEMO ________________________________

____

____

____

PROMPT TITLE: ______________________________

DATE: ______________________ **VERSION:** ____________

TASK / GOAL:

________________________ ________________________

________________________ ________________________

CONTEXT & SPECIFICS:

________________________ ________________________

________________________ ________________________

EXAMPLES TO INCLUDE:

Good example:

________________________ ________________________

________________________ ________________________

What to avoid:

________________________ ________________________

________________________ ________________________

STRUCTURE/STEPS:

1. ______________________ 5. ______________________

2. ______________________ 6. ______________________

3. ______________________ 7. ______________________

4. ______________________ 8. ______________________

ACTUAL PROMPT:

____

____

____

RESULTS/NOTES:

What worked: ______________________________________

What didn't: ______________________________________

Refinements for next time:______________________________

____

PROJECT/CONTEXT: ____________________

ROLE / PERSONA:______________________________________

□ Expert consultant
□ Creative writer
□ Data analyst
□ Teacher/educator
□ Other: __________________

TONE CHECKLIST:
□ Professional
□ Casual/friendly
□ Technical
□ Educational
□ Creative

OUTPUT FORMAT:
□ Bullet points
□ Numbered list
□ Paragraphs
□ Table/chart
□ Step-by-step

AUDIENCE: ____

KEY CONSTRAINTS:

____

____

____

MEMO ____

____

____

____

PROMPT TITLE: ______________________

DATE: ______________________ **VERSION:** ______________

TASK / GOAL:

______________________ ______________________

______________________ ______________________

CONTEXT & SPECIFICS:

______________________ ______________________

______________________ ______________________

EXAMPLES TO INCLUDE:

Good example:

______________________ ______________________

______________________ ______________________

What to avoid:

______________________ ______________________

______________________ ______________________

STRUCTURE/STEPS:

1. ______________________
2. ______________________
3. ______________________
4. ______________________
5. ______________________
6. ______________________
7. ______________________
8. ______________________

ACTUAL PROMPT:

____

____

____

RESULTS/NOTES:

What worked: ______________________________

What didn't: ______________________________

Refinements for next time: ______________________

____

PROJECT/CONTEXT: ______________________

ROLE / PERSONA:________________________________

□ Expert consultant
□ Creative writer
□ Data analyst
□ Teacher/educator
□ Other: ________________

TONE CHECKLIST:
□ Professional
□ Casual/friendly
□ Technical
□ Educational
□ Creative

OUTPUT FORMAT:
□ Bullet points
□ Numbered list
□ Paragraphs
□ Table/chart
□ Step-by-step

AUDIENCE: ____

KEY CONSTRAINTS:

____

____

____

MEMO ____

____

____

____

PROMPT TITLE: ______________________________

DATE: ______________________ **VERSION:** ____________

TASK / GOAL:

______________________ ______________________

______________________ ______________________

CONTEXT & SPECIFICS:

______________________ ______________________

______________________ ______________________

EXAMPLES TO INCLUDE:

Good example:

______________________ ______________________

______________________ ______________________

What to avoid:

______________________ ______________________

______________________ ______________________

STRUCTURE/STEPS:

1. ______________________ 5. ______________________
2. ______________________ 6. ______________________
3. ______________________ 7. ______________________
4. ______________________ 8. ______________________

ACTUAL PROMPT:

____

____

____

RESULTS/NOTES:

What worked: ______________________________

What didn't: ______________________________

Refinements for next time: ______________________

____

PROJECT/CONTEXT: ______________________

ROLE / PERSONA:_________________________________

□ Expert consultant
□ Creative writer
□ Data analyst
□ Teacher/educator
□ Other: ____________________

TONE CHECKLIST:
□ Professional
□ Casual/friendly
□ Technical
□ Educational
□ Creative

OUTPUT FORMAT:
□ Bullet points
□ Numbered list
□ Paragraphs
□ Table/chart
□ Step-by-step

AUDIENCE: ______________________________________

KEY CONSTRAINTS:

____

____

____

MEMO ____

____

____

____

PROMPT TITLE: ______________________________

DATE: ______________________ **VERSION:** ______________

TASK / GOAL:

______________________ ______________________

______________________ ______________________

CONTEXT & SPECIFICS:

______________________ ______________________

______________________ ______________________

EXAMPLES TO INCLUDE:

Good example:

______________________ ______________________

______________________ ______________________

What to avoid:

______________________ ______________________

______________________ ______________________

STRUCTURE/STEPS:

1. ______________________
2. ______________________
3. ______________________
4. ______________________
5. ______________________
6. ______________________
7. ______________________
8. ______________________

ACTUAL PROMPT:

____

____

____

RESULTS/NOTES:

What worked: ______________________________

What didn't: ______________________________

Refinements for next time: ______________________

____

PROJECT/CONTEXT: ____________________

ROLE / PERSONA:____________________________________

□ Expert consultant
□ Creative writer
□ Data analyst
□ Teacher/educator
□ Other: __________________

TONE CHECKLIST:
□ Professional
□ Casual/friendly
□ Technical
□ Educational
□ Creative

OUTPUT FORMAT:
□ Bullet points
□ Numbered list
□ Paragraphs
□ Table/chart
□ Step-by-step

AUDIENCE: ____

KEY CONSTRAINTS:

____

____

____

MEMO ____

____

____

____

PROMPT TITLE: ______________________________

DATE: ______________________ **VERSION:** ______________

TASK / GOAL:

__________________________ __________________________

__________________________ __________________________

CONTEXT & SPECIFICS:

__________________________ __________________________

__________________________ __________________________

EXAMPLES TO INCLUDE:

Good example:

__________________________ __________________________

__________________________ __________________________

What to avoid:

__________________________ __________________________

__________________________ __________________________

STRUCTURE/STEPS:

1. ________________________ 5. ________________________

2. ________________________ 6. ________________________

3. ________________________ 7. ________________________

4. ________________________ 8. ________________________

ACTUAL PROMPT:

____

____

____

RESULTS/NOTES:

What worked: ____

What didn't: ____

Refinements for next time: ____________________________

____

PROJECT/CONTEXT: ______________________

ROLE / PERSONA:____

□ Expert consultant
□ Creative writer
□ Data analyst
□ Teacher/educator
□ Other: ____________________

TONE CHECKLIST:
□ Professional
□ Casual/friendly
□ Technical
□ Educational
□ Creative

OUTPUT FORMAT:
□ Bullet points
□ Numbered list
□ Paragraphs
□ Table/chart
□ Step-by-step

AUDIENCE: ____

KEY CONSTRAINTS:

____

____

____

MEMO ____

____

____

____

PROMPT TITLE: ______________________________

DATE: ______________________ **VERSION:** ____________

TASK / GOAL:

______________________ ______________________

______________________ ______________________

CONTEXT & SPECIFICS:

______________________ ______________________

______________________ ______________________

EXAMPLES TO INCLUDE:

Good example:

______________________ ______________________

______________________ ______________________

What to avoid:

______________________ ______________________

______________________ ______________________

STRUCTURE/STEPS:

1. ____________________ 5. ____________________

2. ____________________ 6. ____________________

3. ____________________ 7. ____________________

4. ____________________ 8. ____________________

ACTUAL PROMPT:

____

____

____

RESULTS/NOTES:

What worked: ______________________________

What didn't: ______________________________

Refinements for next time: ______________________________

____

PROJECT/CONTEXT: ____________________

ROLE / PERSONA:____________________________________

□ Expert consultant
□ Creative writer
□ Data analyst
□ Teacher/educator
□ Other: ____________________

TONE CHECKLIST:
□ Professional
□ Casual/friendly
□ Technical
□ Educational
□ Creative

OUTPUT FORMAT:
□ Bullet points
□ Numbered list
□ Paragraphs
□ Table/chart
□ Step-by-step

AUDIENCE: ____

KEY CONSTRAINTS:

____

____

____

MEMO ____

____

____

____

PROMPT TITLE: ______________________________

DATE: ______________________ **VERSION:** ______________

TASK / GOAL:

________________________ ________________________

________________________ ________________________

CONTEXT & SPECIFICS:

________________________ ________________________

________________________ ________________________

EXAMPLES TO INCLUDE:

Good example:

________________________ ________________________

________________________ ________________________

What to avoid:

________________________ ________________________

________________________ ________________________

STRUCTURE/STEPS:

1. ______________________ 5. ______________________
2. ______________________ 6. ______________________
3. ______________________ 7. ______________________
4. ______________________ 8. ______________________

ACTUAL PROMPT:

____

____

____

RESULTS/NOTES:

What worked: ______________________________________

What didn't: ______________________________________

Refinements for next time: __________________________

____

PROJECT/CONTEXT: ______________________

ROLE / PERSONA:________________________________

□ Expert consultant
□ Creative writer
□ Data analyst
□ Teacher/educator
□ Other: ________________

TONE CHECKLIST:
□ Professional
□ Casual/friendly
□ Technical
□ Educational
□ Creative

OUTPUT FORMAT:
□ Bullet points
□ Numbered list
□ Paragraphs
□ Table/chart
□ Step-by-step

AUDIENCE: ____

KEY CONSTRAINTS:

____

____

____

MEMO ____

____

____

____

PROMPT TITLE: ______________________________

DATE: ____________________ **VERSION:** ____________

TASK / GOAL:

____________________ ____________________

____________________ ____________________

CONTEXT & SPECIFICS:

____________________ ____________________

____________________ ____________________

EXAMPLES TO INCLUDE:

Good example:

____________________ ____________________

____________________ ____________________

What to avoid:

____________________ ____________________

____________________ ____________________

STRUCTURE/STEPS:

1. ____________________ 5. ____________________
2. ____________________ 6. ____________________
3. ____________________ 7. ____________________
4. ____________________ 8. ____________________

ACTUAL PROMPT:

____

____

____

RESULTS/NOTES:

What worked: ______________________________

What didn't: ______________________________

Refinements for next time: ____________________

____

PROJECT/CONTEXT: ______________________

ROLE / PERSONA:______________________________________

□ Expert consultant
□ Creative writer
□ Data analyst
□ Teacher/educator
□ Other: ___________________

TONE CHECKLIST:
□ Professional
□ Casual/friendly
□ Technical
□ Educational
□ Creative

OUTPUT FORMAT:
□ Bullet points
□ Numbered list
□ Paragraphs
□ Table/chart
□ Step-by-step

AUDIENCE: ____

KEY CONSTRAINTS:

____

____

____

MEMO ____

____

____

____

PROMPT TITLE: ______________________

DATE: ______________________ **VERSION:** ______________

TASK / GOAL:

______________________ ______________________

______________________ ______________________

CONTEXT & SPECIFICS:

______________________ ______________________

______________________ ______________________

EXAMPLES TO INCLUDE:

Good example:

______________________ ______________________

______________________ ______________________

What to avoid:

______________________ ______________________

______________________ ______________________

STRUCTURE/STEPS:

1. ______________________
2. ______________________
3. ______________________
4. ______________________
5. ______________________
6. ______________________
7. ______________________
8. ______________________

ACTUAL PROMPT:

____

____

____

RESULTS/NOTES:

What worked: ______________________________

What didn't: ______________________________

Refinements for next time: ______________________

____

PROJECT/CONTEXT: ____________________

ROLE / PERSONA:______________________________________

☐ Expert consultant
☐ Creative writer
☐ Data analyst
☐ Teacher/educator
☐ Other: ________________

TONE CHECKLIST:

☐ Professional
☐ Casual/friendly
☐ Technical
☐ Educational
☐ Creative

OUTPUT FORMAT:

☐ Bullet points
☐ Numbered list
☐ Paragraphs
☐ Table/chart
☐ Step-by-step

AUDIENCE: ______________________________________

KEY CONSTRAINTS:

____

____

____

MEMO ______________________________________

____

____

____

PROMPT TITLE: ______________________________

DATE: ______________________ **VERSION:** ______________

TASK / GOAL:

______________________________ ______________________________

______________________________ ______________________________

CONTEXT & SPECIFICS:

______________________________ ______________________________

______________________________ ______________________________

EXAMPLES TO INCLUDE:

Good example:

______________________________ ______________________________

______________________________ ______________________________

What to avoid:

______________________________ ______________________________

______________________________ ______________________________

STRUCTURE/STEPS:

1. ______________________________
2. ______________________________
3. ______________________________
4. ______________________________
5. ______________________________
6. ______________________________
7. ______________________________
8. ______________________________

ACTUAL PROMPT:

____

____

____

RESULTS/NOTES:

What worked: ____

What didn't: ____

Refinements for next time: ________________________________

____

PROJECT/CONTEXT: ____________________

ROLE / PERSONA: ____________________________________

□ Expert consultant
□ Creative writer
□ Data analyst
□ Teacher/educator
□ Other: ____________________

TONE CHECKLIST:
□ Professional
□ Casual/friendly
□ Technical
□ Educational
□ Creative

OUTPUT FORMAT:
□ Bullet points
□ Numbered list
□ Paragraphs
□ Table/chart
□ Step-by-step

AUDIENCE: ____

KEY CONSTRAINTS:

____

____

____

MEMO ____

____

____

____

PROMPT TITLE: ______________________

DATE: ______________________ **VERSION:** ____________

TASK / GOAL:

______________________ ______________________

______________________ ______________________

CONTEXT & SPECIFICS:

______________________ ______________________

______________________ ______________________

EXAMPLES TO INCLUDE:

Good example:

______________________ ______________________

______________________ ______________________

What to avoid:

______________________ ______________________

______________________ ______________________

STRUCTURE/STEPS:

1. ______________________
2. ______________________
3. ______________________
4. ______________________
5. ______________________
6. ______________________
7. ______________________
8. ______________________

ACTUAL PROMPT:

____

____

____

RESULTS/NOTES:

What worked: ______________________________

What didn't: ______________________________

Refinements for next time:______________________

____

PROJECT/CONTEXT: ______________________

ROLE / PERSONA:______________________________________

□ Expert consultant
□ Creative writer
□ Data analyst
□ Teacher/educator
□ Other: ___________________

TONE CHECKLIST:
□ Professional
□ Casual/friendly
□ Technical
□ Educational
□ Creative

OUTPUT FORMAT:
□ Bullet points
□ Numbered list
□ Paragraphs
□ Table/chart
□ Step-by-step

AUDIENCE: ____

KEY CONSTRAINTS:

____

____

____

MEMO ____

____

____

____

PROMPT TITLE: ______________________________

DATE: ______________________ **VERSION:** ____________

TASK / GOAL:

______________________ ______________________

______________________ ______________________

CONTEXT & SPECIFICS:

______________________ ______________________

______________________ ______________________

EXAMPLES TO INCLUDE:

Good example:

______________________ ______________________

______________________ ______________________

What to avoid:

______________________ ______________________

______________________ ______________________

STRUCTURE/STEPS:

1. ______________________
2. ______________________
3. ______________________
4. ______________________
5. ______________________
6. ______________________
7. ______________________
8. ______________________

ACTUAL PROMPT:

____

____

____

RESULTS/NOTES:

What worked: ______________________________

What didn't: ______________________________

Refinements for next time: ______________________

____

PROJECT/CONTEXT: ________________________

ROLE / PERSONA:______________________________________

□ Expert consultant
□ Creative writer
□ Data analyst
□ Teacher/educator
□ Other: ____________________

TONE CHECKLIST:
□ Professional
□ Casual/friendly
□ Technical
□ Educational
□ Creative

OUTPUT FORMAT:
□ Bullet points
□ Numbered list
□ Paragraphs
□ Table/chart
□ Step-by-step

AUDIENCE: ____

KEY CONSTRAINTS:

____

____

____

MEMO ____

____

____

____

PROMPT TITLE: ______________________

DATE: ______________________ **VERSION:** ______________

TASK / GOAL:

______________________ ______________________

______________________ ______________________

CONTEXT & SPECIFICS:

______________________ ______________________

______________________ ______________________

EXAMPLES TO INCLUDE:

Good example:

______________________ ______________________

______________________ ______________________

What to avoid:

______________________ ______________________

______________________ ______________________

STRUCTURE/STEPS:

1. ______________________
2. ______________________
3. ______________________
4. ______________________
5. ______________________
6. ______________________
7. ______________________
8. ______________________

ACTUAL PROMPT:

____

____

____

RESULTS/NOTES:

What worked: ______________________________

What didn't: ______________________________

Refinements for next time: ______________________

____

PROJECT/CONTEXT: ______________________

ROLE / PERSONA:______________________________________

□ Expert consultant
□ Creative writer
□ Data analyst
□ Teacher/educator
□ Other: ___________________

TONE CHECKLIST:
□ Professional
□ Casual/friendly
□ Technical
□ Educational
□ Creative

OUTPUT FORMAT:
□ Bullet points
□ Numbered list
□ Paragraphs
□ Table/chart
□ Step-by-step

AUDIENCE: ____

KEY CONSTRAINTS:

____

____

____

MEMO ____

____

____

____

PROMPT TITLE: ____________________________

DATE: ____________________ **VERSION:** ____________

TASK / GOAL:

____________________ ____________________

____________________ ____________________

CONTEXT & SPECIFICS:

____________________ ____________________

____________________ ____________________

EXAMPLES TO INCLUDE:

Good example:

____________________ ____________________

____________________ ____________________

What to avoid:

____________________ ____________________

____________________ ____________________

STRUCTURE/STEPS:

1. ____________________
2. ____________________
3. ____________________
4. ____________________
5. ____________________
6. ____________________
7. ____________________
8. ____________________

ACTUAL PROMPT:

____

____

____

RESULTS/NOTES:

What worked: ______________________________

What didn't: ______________________________

Refinements for next time: ______________________________

____

PROJECT/CONTEXT: ______________________

ROLE / PERSONA:______________________________________

- □ Expert consultant
- □ Creative writer
- □ Data analyst
- □ Teacher/educator
- □ Other: ___________________

TONE CHECKLIST:

- □ Professional
- □ Casual/friendly
- □ Technical
- □ Educational
- □ Creative

OUTPUT FORMAT:

- □ Bullet points
- □ Numbered list
- □ Paragraphs
- □ Table/chart
- □ Step-by-step

AUDIENCE: ____

KEY CONSTRAINTS:

____

____

____

MEMO ____

____

____

____

PROMPT TITLE: ______________________________

DATE: ________________________ **VERSION:** ______________

TASK / GOAL:

____________________________ ____________________________

____________________________ ____________________________

CONTEXT & SPECIFICS:

____________________________ ____________________________

____________________________ ____________________________

EXAMPLES TO INCLUDE:

Good example:

____________________________ ____________________________

____________________________ ____________________________

What to avoid:

____________________________ ____________________________

____________________________ ____________________________

STRUCTURE/STEPS:

1. ________________________ 5. ________________________

2. ________________________ 6. ________________________

3. ________________________ 7. ________________________

4. ________________________ 8. ________________________

ACTUAL PROMPT:

____

____

____

RESULTS/NOTES:

What worked: ____

What didn't: ____

Refinements for next time: ______________________________

____

PROJECT/CONTEXT: ______________________

ROLE / PERSONA:______________________________________

□ Expert consultant
□ Creative writer
□ Data analyst
□ Teacher/educator
□ Other: ___________________

TONE CHECKLIST:
□ Professional
□ Casual/friendly
□ Technical
□ Educational
□ Creative

OUTPUT FORMAT:
□ Bullet points
□ Numbered list
□ Paragraphs
□ Table/chart
□ Step-by-step

AUDIENCE: ____

KEY CONSTRAINTS:

____

____

____

MEMO ____

____

____

____

PROMPT TITLE: ______________________________

DATE: ____________________ **VERSION:** ____________

TASK / GOAL:

______________________ ______________________

______________________ ______________________

CONTEXT & SPECIFICS:

______________________ ______________________

______________________ ______________________

EXAMPLES TO INCLUDE:

Good example:

______________________ ______________________

______________________ ______________________

What to avoid:

______________________ ______________________

______________________ ______________________

STRUCTURE/STEPS:

1. ____________________
2. ____________________
3. ____________________
4. ____________________
5. ____________________
6. ____________________
7. ____________________
8. ____________________

ACTUAL PROMPT:

____

____

____

RESULTS/NOTES:

What worked: ______________________________

What didn't: ______________________________

Refinements for next time: ______________________

____

PROJECT/CONTEXT: ______________________

ROLE / PERSONA:________________________________

□ Expert consultant
□ Creative writer
□ Data analyst
□ Teacher/educator
□ Other: ____________________

TONE CHECKLIST:
□ Professional
□ Casual/friendly
□ Technical
□ Educational
□ Creative

OUTPUT FORMAT:
□ Bullet points
□ Numbered list
□ Paragraphs
□ Table/chart
□ Step-by-step

AUDIENCE: ____

KEY CONSTRAINTS:

____

____

____

MEMO ____

____

____

____

PROMPT TITLE: ______________________

DATE: ______________________ **VERSION:** ______________

TASK / GOAL:

______________________ ______________________

______________________ ______________________

CONTEXT & SPECIFICS:

______________________ ______________________

______________________ ______________________

EXAMPLES TO INCLUDE:

Good example:

______________________ ______________________

______________________ ______________________

What to avoid:

______________________ ______________________

______________________ ______________________

STRUCTURE/STEPS:

1. ______________________
2. ______________________
3. ______________________
4. ______________________
5. ______________________
6. ______________________
7. ______________________
8. ______________________

ACTUAL PROMPT:

____

____

____

RESULTS/NOTES:

What worked: ______________________________

What didn't: ______________________________

Refinements for next time: ______________________

____

PROJECT/CONTEXT: ______________________

ROLE / PERSONA:______________________________________

- □ Expert consultant
- □ Creative writer
- □ Data analyst
- □ Teacher/educator
- □ Other: ____________________

TONE CHECKLIST:

- □ Professional
- □ Casual/friendly
- □ Technical
- □ Educational
- □ Creative

OUTPUT FORMAT:

- □ Bullet points
- □ Numbered list
- □ Paragraphs
- □ Table/chart
- □ Step-by-step

AUDIENCE: ____

KEY CONSTRAINTS:

____

____

____

MEMO ____

____

____

____

PROMPT TITLE: ______________________________

DATE: ______________________________ **VERSION:** ________________

TASK / GOAL:

______________________________ ______________________________

______________________________ ______________________________

CONTEXT & SPECIFICS:

______________________________ ______________________________

______________________________ ______________________________

EXAMPLES TO INCLUDE:

Good example:

______________________________ ______________________________

______________________________ ______________________________

What to avoid:

______________________________ ______________________________

______________________________ ______________________________

STRUCTURE/STEPS:

1. ______________________________
2. ______________________________
3. ______________________________
4. ______________________________
5. ______________________________
6. ______________________________
7. ______________________________
8. ______________________________

ACTUAL PROMPT:

____

____

____

RESULTS/NOTES:

What worked: ____

What didn't: ____

Refinements for next time: ________________________________

____

PROJECT/CONTEXT: ______________________

ROLE / PERSONA:______________________________________

- □ Expert consultant
- □ Creative writer
- □ Data analyst
- □ Teacher/educator
- □ Other: __________________

TONE CHECKLIST:

- □ Professional
- □ Casual/friendly
- □ Technical
- □ Educational
- □ Creative

OUTPUT FORMAT:

- □ Bullet points
- □ Numbered list
- □ Paragraphs
- □ Table/chart
- □ Step-by-step

AUDIENCE: ____

KEY CONSTRAINTS:

____

____

____

MEMO ____

____

____

____

PROMPT TITLE: ______________________________

DATE: ______________________ **VERSION:** ______________

TASK / GOAL:

__________________________ __________________________

__________________________ __________________________

CONTEXT & SPECIFICS:

__________________________ __________________________

__________________________ __________________________

EXAMPLES TO INCLUDE:

Good example:

__________________________ __________________________

__________________________ __________________________

What to avoid:

__________________________ __________________________

__________________________ __________________________

STRUCTURE/STEPS:

1. ________________________ 5. ________________________

2. ________________________ 6. ________________________

3. ________________________ 7. ________________________

4. ________________________ 8. ________________________

ACTUAL PROMPT:

____

____

____

RESULTS/NOTES:

What worked: ______________________________________

What didn't: ______________________________________

Refinements for next time: __________________________

____

PROJECT/CONTEXT: ____________________

ROLE / PERSONA:____

- □ Expert consultant
- □ Creative writer
- □ Data analyst
- □ Teacher/educator
- □ Other: ____________________

TONE CHECKLIST:

- □ Professional
- □ Casual/friendly
- □ Technical
- □ Educational
- □ Creative

OUTPUT FORMAT:

- □ Bullet points
- □ Numbered list
- □ Paragraphs
- □ Table/chart
- □ Step-by-step

AUDIENCE: ____

KEY CONSTRAINTS:

____

____

____

MEMO ____

____

____

____

PROMPT TITLE: ______________________________

DATE: ______________________ **VERSION:** ______________

TASK / GOAL:

______________________________ ______________________________

______________________________ ______________________________

CONTEXT & SPECIFICS:

______________________________ ______________________________

______________________________ ______________________________

EXAMPLES TO INCLUDE:

Good example:

______________________________ ______________________________

______________________________ ______________________________

What to avoid:

______________________________ ______________________________

______________________________ ______________________________

STRUCTURE/STEPS:

1. ______________________________
2. ______________________________
3. ______________________________
4. ______________________________
5. ______________________________
6. ______________________________
7. ______________________________
8. ______________________________

ACTUAL PROMPT:

____

____

____

RESULTS/NOTES:

What worked: ____

What didn't: ____

Refinements for next time: ________________________________

____

PROJECT/CONTEXT: ________________________

ROLE / PERSONA:____

□ Expert consultant
□ Creative writer
□ Data analyst
□ Teacher/educator
□ Other: ____________________

TONE CHECKLIST:
□ Professional
□ Casual/friendly
□ Technical
□ Educational
□ Creative

OUTPUT FORMAT:
□ Bullet points
□ Numbered list
□ Paragraphs
□ Table/chart
□ Step-by-step

AUDIENCE: ____

KEY CONSTRAINTS:

____

____

____

MEMO ____

____

____

____

PROMPT TITLE: ______________________

DATE: ______________________ **VERSION:** ______________

TASK / GOAL:

______________________ ______________________

______________________ ______________________

CONTEXT & SPECIFICS:

______________________ ______________________

______________________ ______________________

EXAMPLES TO INCLUDE:

Good example:

______________________ ______________________

______________________ ______________________

What to avoid:

______________________ ______________________

______________________ ______________________

STRUCTURE/STEPS:

1. ______________________
2. ______________________
3. ______________________
4. ______________________
5. ______________________
6. ______________________
7. ______________________
8. ______________________

ACTUAL PROMPT:

____

____

____

RESULTS/NOTES:

What worked: ______________________________

What didn't: ______________________________

Refinements for next time: ______________________

____

MAKE A DIFFERENCE WITH YOUR REVIEW

A Tiny Favor that Creates Big Karma

Studies suggest generous people live longer, smile more, and occasionally find twenty dollars in old coat pockets. I cannot promise that will happen, but leaving a review gets you close on the good karma scale.

Would you help a complete stranger and expect nothing in return?

This stranger is basically you a few steps back. Curious. Hopeful. Trying to understand AI without sinking in the digital deep end. This journal was created to help you and them learn faster and feel confident with their prompts.

Your review helps someone decide if this journal is right for their learning journey.

A review takes less than a minute but can help:

- a small business stay afloat
- an entrepreneur support a family
- an employee shine at work
- a client improve their life
- a beginner feel brave

If you want that warm feeling of doing something good today, scan the QR code, or enter the Bit.ly code in your browser, and leave your review.

Amazon Review Link
bit.ly/4q345CT

PS - Be on the lookout for more journals coming soon for your Tweens and Teens. Since young people often know more about technology than adults, we might as well help them use it wisely.

bitly

Scan me

Your Bonus Digital Journal Awaits

As a thank-you for choosing this paperback journal, you're invited to enjoy a complimentary digital version designed to help you write, reflect, and plan wherever you are.

- Use it on your tablet, computer, or preferred note-taking app.
- Keep journaling while traveling, commuting, or relaxing at home.
- Revisit prompts and pages anytime for extra inspiration.

ow to claim your bonus:

1. Type the link below into your browser.
2. Download your bonus Digital Journal.
3. Save a Master to your device.
4. Start journaling anytime and create multiple journals as needed.

https://dawnoneal.com/download-journal-for-adults/

hank you for being part of this journaling journey. May these pages bring you clarity, creativity, and moments of peace.

Dawn O'Neal
AUTHOR

Made in the USA
Monee, IL
04 November 2026

67381903R00069